Learn the Times Tables While Having Fun

A useful volume for quickly learning math times tables while having fun!

A. Derran

Copyright © 2024

Learn the Times Tables While Having Fun !!!

1.Introduction to the World of Multiplication Tables

Hello! Today we're starting an exciting adventure into the world of **multiplication tables**. Multiplication tables are a bit like magical formulas that help us do math faster. You'll get to know them and use them every day, just like a superhero uses their powers!

1.1 **What are multiplication tables?**

Multiplication tables are series of numbers that teach us how to multiply. Multiplying is like adding the same number many times! For example, if you want to know how much **3 x 4** is, instead of adding **3 + 3 + 3 + 3** (which can take a while), with the multiplication table you can immediately know that the answer is **12**!

Each number has its own table, from the 2 times table all the way up to the 10 times table (and even beyond, if you're really curious!). It's like having a shortcut for doing math!

1.2 **Why are they important?**

Multiplication tables are super important because they help us do calculations much faster. When you know the tables, you can solve math problems without having to think too hard!

Imagine you're in a store and need to buy 3 toys that

cost 5 coins each. If you know the 5 times table, you'll know right away that **3 x 5 = 15**! This way, you can quickly count the coins and know exactly how much to spend.

Also, knowing your multiplication tables well will help you with other things in the future, like division (which is kind of like the opposite of multiplication) or solving more difficult math problems. So, learning the tables is like building a strong foundation for all the other things you'll learn!

Now, are you ready to discover more about this world? Let's take the first step together!

Chapter 1: Learning the 1 Times Table – A Super Easy and Fun Game!

The **1 times table** is one of the easiest of all the tables, but it's also very useful! Let me tell you why: every time you multiply a number by 1, the result is always the same number. Easy, right? Let's see how to learn it in a fun way and how you can memorize it without any effort!

1. Understanding the 1 Times Table: The Magic Power of the Number 1
The rule is super simple: **any number multiplied by 1 stays the same number**!

Here are some examples to help you understand:
1 x 1 = 1
2 x 1 = 2
3 x 1 = 3
4 x 1 = 4
5 x 1 = 5
6 x 1 = 6
7 x 1 = 7
8 x 1 = 8
9 x 1 = 9
10 x 1 = 10

As you can see, the number doesn't change when you multiply it by 1. It's like the number 1 is a magical mirror that reflects the number and keeps it the same!

2. **How to Memorize It in a Fun Way**

Now, let's see how to memorize this table by playing some games!

2.1 **The Magic Mirror Game**
Imagine that the number 1 is a magical mirror. You put a number in front of it (like 3), and the mirror reflects the same number: **3 x 1 = 3**.

Try playing this game with your fingers:
- Hold up one hand and put up as many fingers as the number you want to multiply (for example, 4).
- Then imagine touching an "invisible mirror" that is the number 1. What do you see reflected? The number 4!

This mirror game will help you remember that whenever you multiply by 1, the number stays the same!

2.2 **The 1 Times Table Song**
Songs are a great way to remember things! You can make up a little rhyme for the 1 times table. Here's an example:

🎵 🎶 **One, one, it's really great,**
multiply by anything, and nothing will change!
Three times one is always three,
four times one stays as it should be! 🎶 🎵

You can sing this little song while doing math, and it'll

be easier to remember the answers!

2.3 **Draw and Color**
Another fun way to learn the times table is by using drawings. Take a piece of paper and draw the number 1 big in the center, like it's a superhero. Then around it, draw the numbers you want to multiply (2, 3, 4...) and write the result next to them.

For example:
- Draw a 2 and write next to it: **2 x 1 = 2**
- Draw a 5 and write: **5 x 1 = 5**

You can color everything and hang the drawing in your room. That way, every time you look at it, you'll remember the times table!

3. **Quick Practice Game!**

Once you've learned the times table, you can have fun with a timed game! Ask a friend or an adult to ask you questions from the 1 times table, and try to answer as quickly as you can. For example, they ask: **7 x 1?** and you answer immediately **7!**. See how many you can answer in one minute!

Conclusion: The 1 Times Table Is the First Step!
Learning the 1 times table is super easy and fun. With the mirror game, the little song, and the drawings, you'll remember it without any trouble. Now you're

ready to take on the other times tables, but don't forget that every journey starts with a small step, and the 1 times table is that very first step!

Chapter 2
Learning the 2 Times Table: A Fun Adventure!

Hello! Today we're going to learn the **2 times table** in a super fun way! The 2 times table is special because it's like making **jumps** forward by two numbers each time! Sounds almost like a game, right? Get ready, because you'll become a champion of the 2 times table in no time!

1. What is the 2 Times Table?

The 2 times table simply means multiplying a number by 2. And every time you multiply by 2, you're adding that number to itself twice! For example:

- **2 x 1** is like saying **1 + 1**, which equals **2**!
- **2 x 2** is like saying **2 + 2**, which equals **4**!

See? Easy, right? Now, here's the whole 2 times table:

- 2 x 1 = 2
- 2 x 2 = 4
- 2 x 3 = 6
- 2 x 4 = 8
- 2 x 5 = 10
- 2 x 6 = 12
- 2 x 7 = 14
- 2 x 8 = 16
- 2 x 9 = 18

- 2 x 10 = 20

Let's take a leap forward two numbers at a time!

2. Learning the 2 Times Table in a Fun Way

Now that you know how to multiply by 2, let's explore some games and tricks to remember it in a fun way!

2.1 The Jumping Game

Imagine you're a bunny hopping through a field! Every time you multiply by 2, you make a jump forward, counting by 2's:

- Jump once: 2 x 1 = 2
- Jump twice: 2 x 2 = 4
- Jump three times: 2 x 3 = 6

You can play this game at home too: place cards or sheets of paper with the numbers 2, 4, 6, 8… on the floor, and hop from one number to the next. With each jump, say the number out loud!

2.2 2 Times Table Song

Songs are great for helping us remember things! Here's a little song for the 2 times table that you can sing while doing your sums:

? Two times one is two, let's jump!
Two times two is four, let's keep going!
Two times three is six, see how high we go!
Two times four is eight, let's make a great jump! ?

Try making up your own song or continue with the rest of the table! Singing will help you remember more easily.

2.3 Counting Objects in Pairs

Gather some of your favorite objects, like toys, candies, or pencils, and group them in pairs! For example, if you take **3 pairs of candies**, you're multiplying **2 x 3**, so you'll have **6 candies**.

Here are some other things you can do:

- **2 x 1**: take one pair of toy cars, and you have **2 cars**.
- **2 x 3**: take 3 pairs of pencils and count: **2, 4, 6 pencils**!
- **2 x 5**: take 5 pairs of stickers, and you'll have **10 stickers**.

Playing with objects helps you better understand the concept of multiplying by 2.

2.4 Draw and Color the Times Table

Grab a piece of paper and draw a staircase with 10 steps. Each step represents one result of the 2 times table! Write the numbers on each step: **2, 4, 6, 8... up to 20**. Then color each step in a different color.

Each time you climb the steps, repeat the results of the times table. This will help you visualize the numbers and remember them better!

2.5 The Timed Game

Now that you've learned the 2 times table, let's have a race! Ask a friend or an adult to quiz you on the 2 times table, and try to answer as quickly as possible.

For example, if they ask: **2 x 7?**, you quickly reply **14!**. See how many answers you can give in one minute!

3. Reviewing and Becoming a 2 Times Table Champion

Once you've learned the 2 times table, you can keep practicing it every day. Here are some ways to do it:

- Play the jumping game with numbers in the garden or at home.
- Sing the 2 times table song while taking a walk or playing.
- Draw more staircases or invent new games!

The 2 times table is really fun and easy to learn! With jumps, songs, drawings, and games, you'll become a true champion. Remember, each time you multiply by 2, you're taking a jump forward by two numbers! Are you ready to discover more times tables?

Chapter 3
Learning the 3 Times Table: A Game of Magic and Numbers!

Hello, little number explorer! Today we're venturing into the magical world of the 3 times table. This is a truly special times table because it's like jumping forward, but this time each jump is by three numbers! And guess what? It can be super fun to learn! Are you ready? Let's go on this journey together!

1. What is the 3 Times Table?

The 3 times table is like a magic formula that helps us multiply any number by 3. Each time you multiply by 3, it's like adding that number three times! Let's do some examples right away:

- **3 x 1 = 3** (it's like saying **1 + 1 + 1**)
- **3 x 2 = 6** (it's like saying **2 + 2 + 2**)

Here's the whole 3 times table:

- 3 x 1 = 3
- 3 x 2 = 6
- 3 x 3 = 9
- 3 x 4 = 12
- 3 x 5 = 15
- 3 x 6 = 18
- 3 x 7 = 21
- 3 x 8 = 24

- 3 x 9 = 27
- 3 x 10 = 30

Every time you jump forward by 3 numbers. Easy, right? Now let's see how to learn it in a super fun way!

2. How to Learn the 3 Times Table in a Fun Way

Now that you've seen how it works, it's time to discover some secrets to learning it while playing and having fun!

2.1 The Magic Jumping Game

Imagine you're a wizard with a magic cape. Each time you multiply by 3, you make a magical jump forward by three numbers. Try counting while you jump:

- Make the first jump: **3 x 1 = 3**
- Make the second jump: **3 x 2 = 6**
- Make the third jump: **3 x 3 = 9**

You can play this game at home too: take some small pieces of paper and write the numbers of the 3 times table on them (3, 6, 9, 12…). Place them on the floor, one next to the other, and try jumping from one number to the next, saying the result out loud with each jump!

2.2 3 Times Table Song

Songs really help us remember things! Here's a little song for the 3 times table that you can sing while you do your sums:

**? Three times one is three, let's jump up high!
Three times two is six, now let's fly!**

**Three times three is nine, we all jump in a row,
Three times four is twelve, now let's take off and go!** ?

Try singing this song, adding the numbers up to **3 x 10 = 30**. Singing makes everything easier and more fun!

2.3 The Ball Game

Take some balls or any other objects you can count. For each multiplication, count the right number of balls and group them in threes.

For example:

- For **3 x 2**, take 2 groups of 3 balls and count up to 6.
- For **3 x 4**, take 4 groups of 3 balls and count up to 12.

Using balls will help you see and touch the numbers, making multiplication easier to understand!

2.4 Draw and Color

Take a sheet of paper and draw a big carousel with 10 seats! On each seat, write one result of the 3 times table. For example:

- On the first seat, write **3 (3 x 1 = 3)**
- On the second seat, write **6 (3 x 2 = 6)**
- On the third seat, write **9 (3 x 3 = 9)**

Then color each seat in a different color. When you're done, you'll have a colorful carousel that will help you remember all the results!

2.5 Speed Race

Now that you've learned the 3 times table, you can challenge yourself with a little race! Ask a friend or an adult to quiz you, and try to answer as fast as possible. For example, if they ask **3 x 7?**, you answer right away **21!**.

Have a race against the clock and see how many answers you can give in one minute!

3. Reviewing the 3 Times Table and Becoming a Super Champion

Now that you know the tricks, it's important to review every day! You can do it in lots of fun ways:

- Play with your magic jumps.
- Sing the 3 times table song while you're out and about.
- Use balls or toys to count the numbers.

Every time you review, you'll get better and better!

Conclusion: The 3 Times Table is Easy and Fun!

The 3 times table is like a game where you make magical jumps by 3 numbers at a time. With jumps, songs, games, and drawings, learning becomes a fun adventure! Now, are you ready to discover more times tables? Have fun! ?

Chapter 4: Learning the 4 Times Table: A Team Game!

Hello! Today, we're diving into the fantastic world of the 4 times table. Learning it is like being part of a special team that always moves forward by fours. Are you ready to become a champion of Team 4? With games, jumps, and lots of fun ideas, the 4 times table will be super easy to remember!

1. What is the 4 Times Table?

The 4 times table is like a magic trick that helps us multiply any number by 4. Each time you multiply by 4, you're adding that number four times. Let's look at a few examples:

- **4 x 1 = 4** (it's like saying **1 + 1 + 1 + 1 = 4**)
- **4 x 2 = 8** (it's like saying **2 + 2 + 2 + 2 = 8**)

Now, here's the whole 4 times table:

- 4 x 1 = 4
- 4 x 2 = 8
- 4 x 3 = 12
- 4 x 4 = 16
- 4 x 5 = 20
- 4 x 6 = 24
- 4 x 7 = 28

- 4 x 8 = 32
- 4 x 9 = 36
- 4 x 10 = 40

Every time you multiply by 4, you take a big step forward by 4 numbers! Now, let's see how to learn it in a fun way!

2. How to Learn the 4 Times Table in a Fun Way

Learning doesn't have to be boring—in fact, with these games and tricks, the 4 times table will become your favorite game!

2.1 The Giant Steps Game

Imagine you're a giant walking with enormous steps! Each step is 4 numbers long.

- The first step is **4 x 1 = 4**.
- Then take another giant step: **4 x 2 = 8**.
- Another one: **4 x 3 = 12**!

You can try doing this for real! Find a big space and take giant steps, counting out loud each result of the 4 times table. Every time you take a step, you jump forward by 4 numbers!

2.2 4 Times Table Song

Singing is one of the best ways to memorize times tables! Here's a song you can sing while learning the 4 times table:

♪ **Four times one is four, I know!**
Four times two is eight, let's jump high!
Four times three, twelve is here,
Four times four, sixteen, cheer! ♪

You can keep singing all the way up to **4 x 10 = 40**. Do some movements while you sing, so your body remembers the numbers better!

2.3 Building Block Game

If you have building blocks (like Legos), this game is perfect for you! Take some blocks and group them in fours. Then, for each multiplication in the 4 times table, build a tower with the right number of blocks.

Here's how:

- **4 x 1**: Build a tower with 4 blocks.
- **4 x 2**: Build a tower with 8 blocks (4 + 4).
- **4 x 3**: Build a tower with 12 blocks (4 + 4 + 4).

Keep going all the way to **4 x 10 = 40**! In the end, you'll have a super colorful tower that will help you remember the 4 times table.

2.4 Drawing and Coloring the 4 Times Table

Take a piece of paper and draw 10 circles in a row. These represent the results of the 4 times table. Inside each circle, write a result and color the circle with your favorite color!

For example:

- In the first circle, write **4 (4 x 1 = 4)** and color it blue.
- In the second circle, write **8 (4 x 2 = 8)** and color it red.

When you're finished, you'll have a row of colorful circles representing all the results of the 4 times table! Hang the paper in your room to review it every day.

2.5 Speed Race with a Friend

Now that you know the 4 times table, you can have fun with a little race! Ask a friend or an adult to quiz you. For example, if they ask, **4 x 7?**, answer right away **28!**.

Try to respond as quickly as possible. See how many answers you can give in one minute. You can even challenge your friends to see who's the fastest!

3. Reviewing and Having Fun Every Day

Now that you've learned the 4 times table, reviewing it every day will become a game! Here are some ways to do it:

- Do the giant steps game every time you walk from one room to another.
- Sing the song while you're in the car or while you play.
- Build towers with your blocks and remember the numbers.

By reviewing every day, you'll become a champion in no time!

Conclusion

The 4 times table is like a team game, where you take giant steps and count by fours! With games, songs, building blocks, and drawings, learning it will be easy and fun. Now you're ready to become a true times table expert!

Chapter 5: Learning the 5 Times Table: A Fun Journey!

Hello, little explorer! Today, we're setting off on a fantastic journey into the world of the 5 times table. Do you know why it's so special? Because multiplying by 5 is like counting with your fingers or looking at the clock! And with so many fun games, learning this times table will be an adventure! Are you ready? Let's go!

1. What is the 5 Times Table?

The 5 times table is like a perfect rhythm: every time you multiply by 5, you jump forward by 5 numbers! Let's see how it works:

- 5 x 1 = 5
- 5 x 2 = 10
- 5 x 3 = 15
- 5 x 4 = 20
- 5 x 5 = 25
- 5 x 6 = 30
- 5 x 7 = 35
- 5 x 8 = 40
- 5 x 9 = 45
- 5 x 10 = 50

As you can see, every result ends in either 0 or 5! This

is a trick that will help you remember it easily. Now, let's learn the 5 times table in a super fun way!

2. How to Learn the 5 Times Table in a Fun Way

Now that you know how it works, let's discover the games and activities that will make learning easy and fun!

2.1 The Hands Game

Have you noticed that each hand has 5 fingers? This game is based on counting with your fingers!

- Raise 1 hand: **5 x 1 = 5** (count the fingers!).
- Raise 2 hands: **5 x 2 = 10** (you have 10 fingers in total!).

Try to keep going with more friends or family members: every time you add a hand, you add 5! You can also play with a friend: one of you raises their hands, and the other shouts out the answer!

2.2 The 5 Times Table Song

Learning through singing is always easier! Here's a song you can sing while reviewing the 5 times table:

♪ **Five times one is five, let's go!**
Five times two is ten, so much fun!
Five times three, fifteen's the key,
Five times four, twenty, yippee! ♪

Try singing it by adding the numbers all the way to **5 x**

10 = 50. You can even create your own version or add some dance moves!

2.3 The Clock Game

Did you know that every time the clock moves by one number, it's 5 minutes? You can use a clock to learn the 5 times table! Here's how:

- Look at the clock and point the hand to each number. For example, when the hand points to 1, it's 5 minutes (just like **5 x 1 = 5**).
- When the hand points to 2, it's 10 minutes (just like **5 x 2 = 10**).

This is a super fun way to see the 5 times table in action! Every number on the clock helps you review the results.

2.4 Play with Coins

Grab some 5-cent coins (if you have them) and use them to practice! Each coin represents the number 5.

- Take one coin: **5 x 1 = 5**.
- Take two coins: **5 x 2 = 10**.

Keep adding coins and count how much you have with 3, 4, or 5 coins. It's like a little treasure that keeps growing! You can even pretend you're in a store and give the right "change" using the results of the 5 times table.

2.5 Draw and Color the 5 Times Table

Take a piece of paper and draw 10 circles in a row. Inside each circle, write a result of the 5 times table. Then, color each circle in a different color.

- In the first circle, write **5** and color it blue.
- In the second circle, write **10** and color it red.
- In the third circle, write **15**, and so on, all the way to 50!

When you're done, you'll have a series of colorful circles representing all the results of the 5 times table. Every time you look at them, you'll remember the numbers!

2.6 Speed Race with a Ball

Here's a fun game you can play with a ball! Ask a friend or an adult to quiz you on the 5 times table. Each time you answer correctly, you get to throw the ball.

For example, if they ask you **5 x 4?**, answer quickly **20!** and then take a shot. See how many shots you can make while answering the multiplications!

3. Review Every Day

Now that you've learned all the tricks, it's important to review the 5 times table every day! You can do it by playing, singing, or even drawing. Here are a few ideas to keep having fun while you review:

- Play the Hands Game with your friends or family.
- Sing the song while you walk or get ready for school.
- Play with the clock every time you check the time!

Every time you review, you'll get better and faster!

Conclusion

The 5 times table is like a rhythmic game, where you can count using your hands, the clock, or even coins. With games, songs, and fun activities, learning it will be easy and enjoyable! Are you ready to become a 5 times table expert? Have fun!

Chapter 6: Learning the 6 Times Table: A Fun Number Adventure!

Hello, number explorer! Today we're diving into the 6 times table, a table that will show you just how fun playing with numbers can be. Do you know why it's special? Because multiplying by 6 is like making huge jumps and reaching new milestones every time. With games, songs, and creative activities, learning the 6 times table will be a fun adventure! Ready to begin?

1. What is the 6 Times Table?

The 6 times table helps you multiply numbers by 6. Every time you multiply by 6, you're adding that number six times! Let's look at a few examples:

- **6 x 1 = 6** (it's like saying 1 + 1 + 1 + 1 + 1 + 1 = 6)
- **6 x 2 = 12** (it's like saying 2 + 2 + 2 + 2 + 2 + 2 = 12)

Here's the full 6 times table:

- 6 x 1 = 6
- 6 x 2 = 12
- 6 x 3 = 18
- 6 x 4 = 24
- 6 x 5 = 30
- 6 x 6 = 36

- 6 x 7 = 42
- 6 x 8 = 48
- 6 x 9 = 54
- 6 x 10 = 60

Each time you multiply by 6, you move forward by six steps! Now, let's see how to learn the 6 times table while playing and having fun.

2. How to Learn the 6 Times Table in a Fun Way

Learning doesn't have to be boring—in fact, it can be exciting! With these tricks and games, the 6 times table will become your favorite game!

2.1 The Giant Jumps Game

Imagine you're a superhero making giant leaps. Each time you multiply by 6, you take a huge jump! Try doing it for real:

- Take the first jump: **6 x 1 = 6**.
- Take the second jump: **6 x 2 = 12**.
- Take the third jump: **6 x 3 = 18**.

Find a big space and try making real jumps from one number to the next. Each time you jump, you have to say the multiplication result. It's a fun way to use both your body and mind together!

2.2 The 6 Times Table Song

Singing makes everything easier! Here's a song you

can sing while reviewing the 6 times table:

♪ **Six times one is six, you know,
Six times two, twelve, here we go!
Six times three, eighteen is the key,
Six times four, twenty-four will be!** ♪

You can continue the song up to **6 x 10 = 60**. Sing the song while dancing or moving around, so your body helps you remember the numbers!

2.3 Build the 6 Tower

Grab some building blocks or anything you can stack, like bottle caps or small objects. Build a tower that represents the 6 times table. For each multiplication, add a piece to the tower:

- **6 x 1 = 6**: Take 6 blocks and build the first part of the tower.
- **6 x 2 = 12**: Add 6 more blocks, so the tower grows to 12.
- **6 x 3 = 18**: Add 6 more blocks, and the tower rises to 18.

Continue up to **6 x 10 = 60**. In the end, you'll have a super tall tower that represents all the results of the 6 times table!

2.4 Dice Throwing Game

For this game, you need a die. Each time you roll the die, you have to multiply the number by 6.

- If you roll a 3, you have to say **6 x 3 = 18**.
- If you roll a 5, you have to say **6 x 5 = 30**.

You can play alone or with a friend. If you're playing with someone, whoever makes fewer mistakes wins! It's a fun way to practice while throwing the dice!

2.5 Draw the 6 Ladder

Take a piece of paper and draw a big ladder with 10 steps. On each step, write one of the results of the 6 times table.

- On the first step, write **6** (6 x 1 = 6).
- On the second step, write **12** (6 x 2 = 12).

Keep going until the 10th step, where you'll write **60** (6 x 10 = 60). Color each step in a different color. When you're done, you'll have a rainbow ladder that will help you remember the 6 times table every time you look at it!

2.6 The Magic Ball Game

Grab some balls or small objects. Each ball represents a group of 6. Now, for every multiplication in the 6 times table, add the right number of balls:

- For **6 x 1**, take 1 group of 6 balls.
- For **6 x 2**, take 2 groups of 6 balls and count up to 12.
- For **6 x 3**, take 3 groups of 6 balls and count up to 18.

Keep going up to **6 x 10 = 60**. It's a fun way to see and touch the numbers while learning them!

3. Review Every Day

Now that you know all the tricks for learning the 6 times table, it's important to review it every day. Here are some ways to keep reviewing:

- Play the Giant Jumps Game while walking from one room to another.
- Sing the 6 times table song while getting ready or during a car ride.
- Play with the dice or build the 6 tower whenever you have a few minutes.

Reviewing every day makes everything easier, and soon you'll be a true champion of the 6 times table!

Conclusion

The 6 times table is like an adventure full of jumps, songs, and games. With these fun tricks, learning it will be a breeze! Remember to review often, and soon you'll be a real expert in the 6 times table. Have fun!

Chapter 7: Learning the 7 Times Table: A Super Fun Adventure!

Hey, number explorer, are you ready for a new mission? Today, we're diving into the 7 times table, a special table that may seem a bit tricky at first, but with the right games and tricks, it will become easy and fun. Learning the 7 times table is like going on a journey full of surprises and discoveries. Together, we'll find the most exciting ways to remember every number! Ready? Let's go!

1. What is the 7 Times Table?

The 7 times table is a way to multiply any number by 7. Each time you multiply by 7, you take a big leap, adding 7 numbers at a time. Here's how it works:

- 7 x 1 = 7
- 7 x 2 = 14
- 7 x 3 = 21
- 7 x 4 = 28
- 7 x 5 = 35
- 7 x 6 = 42
- 7 x 7 = 49
- 7 x 8 = 56
- 7 x 9 = 63
- 7 x 10 = 70

Now that you know all the answers, I'll show you how to make them easy and fun to remember!

2. How to Learn the 7 Times Table in a Fun Way

Learning is always better when we do it while playing! Here are some tricks and activities that will make learning the 7 times table a joyful walk in the park.

2.1 The Giant Steps Game

Ready to move around a little? Every time you multiply by 7, take a big leap or a giant step!

- Take one step: **7 x 1 = 7**.
- Take another giant step: **7 x 2 = 14**.

Keep taking giant steps, saying the multiplication result out loud. You can play this game outside or even indoors—just make sure you have enough space to jump! Every time you reach a new result, imagine you're jumping from one mountain to another!

2.2 The 7 Times Table Song

Singing makes everything easier, right? Here's a song you can sing while repeating the 7 times table. Add a happy rhythm and maybe even some dance moves!

♪ **Seven times one is seven! Olé!**
Seven times two is fourteen, hooray!
Seven times three, twenty-one is here,
Seven times four, twenty-eight is near! ♪

Keep going until **7 x 10 = 70**, adding your favorite rhythm. You can sing this song while walking to school or getting ready in the morning. The more you sing, the easier it will be to remember the numbers!

2.3 The 7 Puzzle Game

Create your own 7 times table puzzle! Take a piece of paper and draw puzzle pieces. On each piece, write a multiplication from the 7 times table, like:

- 7 x 1 = 7
- 7 x 2 = 14

Then cut out each piece and mix them up. Now, you need to solve the puzzle by matching each multiplication to the correct answer! This game will help you review and have fun at the same time. You can play alone or challenge a friend to see who can complete the puzzle faster!

2.4 The Magic Dice Game

Have you ever played with dice? Here's how you can use them to review the 7 times table!
Roll a die: the number you get is the one you need to multiply by 7.

- If you roll a 4, you need to say: **7 x 4 = 28**.
- If you roll a 6, say: **7 x 6 = 42**.

Each time you roll the die, you have to say the result quickly. Play with a friend or your parents and see who answers faster. The more you roll, the better you'll get!

2.5 Draw the 7 Times Table Ladder

Another fun way to learn the 7 times table is by drawing a big ladder with 10 steps, one for each multiplication. On each step, write the result of the 7 times table:

- On the first step, write **7**.
- On the second step, write **14**.
- On the third step, write **21**.

Color each step in a different color. When you reach the top, you'll have reached **7 x 10 = 70**! Each time you look at the colorful ladder, you'll easily remember the numbers.

2.6 Build the 7 Tower with Blocks

Grab some building blocks or small objects like bottle caps or toy bricks. Build a tall tower, adding one block for each multiplication of the 7 times table. For example:

- Place 7 blocks for **7 x 1 = 7**.
- Add 7 more blocks for **7 x 2 = 14**.

Keep adding blocks until **7 x 10 = 70**. When you're done, you'll have a super tall tower that represents all the results of the 7 times table. Try counting them and see how the tower grows taller and taller!

3. Review Every Day with Joy

Now that you've discovered so many games and tricks, it's important to review the 7 times table every day to remember it well. Here are some ideas to make reviewing even more fun:

- Play with your friends and race through the Giant Steps or 7 Puzzle games.
- Sing the 7 times table song while doing other activities like brushing your teeth or tidying up your room.
- Challenge a family member to the Magic Dice Game and see who answers the fastest!

Each day you review, you'll get faster and better, and soon you'll be a true expert of the 7 times table.

The 7 times table isn't so hard when you learn it by playing and having fun! With giant leaps, songs, puzzles, and dice games, you'll become a champion in no time. Remember to review every day, and soon you'll realize just how easy and fun it is to multiply by 7. Ready to become a master of the 7 times table? Have fun and good luck, little number explorer!

Chapter 8: Learning the 8 Times Table: A Magical Journey Through Numbers!

Hello, little number explorer! Today, we're ready for a super fun new challenge: learning the 8 times table! The number 8 might seem big, but I promise that with games, songs, and fun activities, it will soon become your favorite number. Want to find out how? Get your imagination ready and let's start this adventure!

1. What is the 8 Times Table?

The 8 times table helps us multiply numbers by 8. Every time you multiply by 8, you take an even bigger leap than with other times tables because 8 is a big number! Here's how it works:

- 8 x 1 = 8
- 8 x 2 = 16
- 8 x 3 = 24
- 8 x 4 = 32
- 8 x 5 = 40
- 8 x 6 = 48
- 8 x 7 = 56
- 8 x 8 = 64
- 8 x 9 = 72
- 8 x 10 = 80

Now that you know all the answers, it's time to

discover how to learn them easily and have fun!

2. How to Learn the 8 Times Table in a Fun Way

Learning has never been this fun! You'll see that with these games, tricks, and activities, the 8 times table will become a walk in the park.

2.1 The Eight Leaps Game

Imagine you are a magical kangaroo that can make huge leaps, just like the leaps of the number 8! Every time you multiply by 8, take a big leap forward.

- Take one leap: **8 x 1 = 8**.
- Take another leap: **8 x 2 = 16**.

Keep leaping, and each time you land, say the result out loud. This game helps you remember the numbers while also getting some movement and fun. The more you leap, the better you'll get! If you can't jump, you can take big steps or even hop from chair to chair, but always safely!

2.2 The 8 Times Table Song

Singing is one of the easiest ways to memorize something! Here's a special song to help you learn the 8 times table. Add a melody you like and sing along:

? **Eight times one is eight, you see!**
Eight times two, sixteen, whee!
Eight times three, twenty-four is here,

Eight times four, thirty-two is near! ?

You can keep going up to **8 x 10 = 80**! You can sing this song while dancing, getting ready for school, or even during a car ride.

2.3 The 8 Wheel Game

Create your own magical 8 times table wheel! Take a piece of paper and draw a large wheel, like a spinner. Divide it into 10 sections and write a multiplication from the 8 times table on each, such as:

- **8 x 1 = 8**
- **8 x 2 = 16**
- **8 x 3 = 24**

Now cut out a cardboard arrow and attach it to the center of the wheel with a pin or thumbtack. Spin the arrow, and when it stops, say the result of the multiplication it points to! Challenge yourself or your friends: whoever answers fastest wins!

2.4 The Colorful Blocks Game

Another way to learn the 8 times table is by playing with colorful blocks or building bricks. Every time you multiply by 8, add a group of 8 blocks to your creation.

- For **8 x 1 = 8**, take 8 blocks and start building.
- For **8 x 2 = 16**, add another 8 blocks.

Keep building until you reach **8 x 10 = 80**. By the end,

you'll have a super tall tower made of blocks, and each block will represent a multiplication of the 8 times table!

2.5 The Dice Rolling Game

For this game, you'll need two dice! Roll the dice and multiply the number you get by 8. For example:

- If you roll a 5, you must say **8 x 5 = 40**.
- If you roll a 3, you say **8 x 3 = 24**.

You can play with a friend or family member: whoever says the result the fastest wins! It's a simple, fun game that helps you learn without even noticing.

2.6 Build the 8 Staircase

Take a piece of paper and draw a staircase with 10 steps. On each step, write one of the results of the 8 times table:

- On the first step, write **8**.
- On the second step, write **16**.
- On the third step, write **24**.

Color each step a different color. Each time you climb a step, say the result out loud. By the time you reach the top, you'll know the whole 8 times table up to **8 x 10 = 80**!

2.7 The Magic 8 Balls Game

For this game, grab some balls or round objects you

can find at home, like bottle caps or small toys. Every time you multiply a number by 8, grab the right number of balls.

- For **8 x 1 = 8**, take 8 balls.
- For **8 x 2 = 16**, take 16 balls (in groups of 8).

Keep counting and adding the balls until you reach **8 x 10 = 80**. By the end, you'll have lots of little groups of 8 balls that will help you visualize the numbers from the times table.

3. Review Every Day with Fun and Games

Now that you know so many fun ways to learn the 8 times table, it's important to review it every day. Here are some ideas to make it easy:

- Play the Eight Leaps Game as you move around the house.
- Sing the 8 Times Table Song while playing or while in the shower.
- Play the Dice Rolling Game with your friends or family.

Reviewing the 8 times table a little each day will help you remember it effortlessly, and soon you'll be an expert!

Conclusion

Learning the 8 times table is like a magical journey: there are giant leaps, fun songs, and colorful games!

Remember to have fun while you learn, and you'll see that the 8 times table will become one of your favorites. Have fun and keep exploring the world of numbers.

Chapter 9: Learning the 9 Times Table: A Fantastic Adventure Among Numbers!

Hello, little adventurer! Are you ready to discover the magical world of the 9 times table? This times table is full of surprises and fun games that will help you memorize everything without even realizing it. With a bit of creativity and lots of fun, you'll become a true multiplication expert! Ready? Let's go!

1. What is the 9 Times Table?

The 9 times table is a special way to multiply numbers by 9. When we multiply, we are simply making jumps forward. Here's how it works:

9 x 1 = 9
9 x 2 = 18
9 x 3 = 27
9 x 4 = 36
9 x 5 = 45
9 x 6 = 54
9 x 7 = 63
9 x 8 = 72
9 x 9 = 81
9 x 10 = 90

Now that you know the results, I'll show you how to learn them in a fun and creative way!

2. How to Learn the 9 Times Table in a Fun Way

Learning the 9 times table can be a big game! Here are some ideas that will make your journey through numbers super fun.

2.1 The 9 Times Table Song

Singing is a fantastic way to memorize! Here's a little song you can sing while learning the 9 times table. Pick a melody you like, such as a popular song or a lullaby!

? Nine times one is nine, that's how it goes!
Nine times two, eighteen, and it shows!
Nine times three, twenty-seven with glee,
Nine times four, thirty-six, come sing with me! ?

Keep singing up to 9 x 10 = 90. Every time you sing, you'll remember the numbers more easily!

2.2 The 9 Jumps Game

Ready for some movement? This game will have you jumping like a kangaroo! Every time you say a multiplication result, take a jump!

Jump for 9 x 1 = 9.
Take another jump for 9 x 2 = 18.
Keep jumping all the way to 9 x 10 = 90!

You can do this outside or even indoors, as long as you have enough space to jump. Every jump will help you remember the results!

2.3 The 9 Times Table Ball Game

Gather some balls or small objects (like bottle caps or building blocks). Each time you multiply a number by 9, take the correct number of balls.

For 9 x 1 = 9, take 9 balls.
For 9 x 2 = 18, take 18 balls (you can do this in groups of 9).
Continue up to 9 x 10 = 90.

By the end, you'll have lots of groups of balls that will help you visualize and remember the 9 times table!

2.4 The 9 Times Table Puzzle Game

Draw a large puzzle on a piece of paper. Each puzzle piece represents a multiplication from the 9 times table. For example:

Draw one piece with 9 x 1 = 9.
Another with 9 x 2 = 18.
And so on up to 9 x 10 = 90.

Cut out the pieces and mix them up. Now, your goal is to put the puzzle back together by matching the results with the correct pieces! This will help you remember the numbers while having fun.

2.5 The Magical 9 Ladder

Take a sheet of paper and draw a ladder with 10 steps. On each step, write one of the results from the 9 times table:

On the first step, write 9.
On the second step, write 18.
On the third step, write 27.

Every time you climb a step, say the result out loud. When you reach the top, you'll have reviewed the entire 9 times table!

2.6 The Magic Dice Game

For this game, take two dice. Roll the dice and multiply the number you get by 9. For example:

If you roll a 5, you must say 9 x 5 = 45.
If you roll a 3, say 9 x 3 = 27.

You can challenge a friend to see who can answer the fastest!

2.7 The 9 Times Table Color Game

Grab some crayons or colored pencils. Draw a big cake or pizza divided into 10 slices. On each slice, write a result from the 9 times table:

First slice: 9
Second slice: 18
And so on, up to 90.

Then color each slice with a different color. Every time you look at your colorful cake, you'll easily remember the 9 times table!

3. Review Every Day with Joy

Now that you know so many games and tricks, it's important to review the 9 times table every day. Here are some ideas:

Jump and sing the 9 times table song while doing your homework.
Play the 9 Times Table Ball Game with a friend.
Do the 9 Puzzle Game and challenge yourself to complete it faster each time.

Reviewing the 9 times table a little each day will help you remember it effortlessly. You'll become a real champion!

Conclusion

Learning the 9 times table is a journey full of adventures and fun! With songs, games, jumps, and colors, this times table will become one of your favorites. Have fun, little number explorer, and remember: math is a great adventure!

Chapter 10: Learning the 10 Times Table Easily!

Learning the 10 Times Table: A Fun Journey Through Numbers!

Hello, champion! Today we're going on an adventure into the fantastic world of the 10 times table. Get ready to discover that learning numbers can be a piece of cake! The 10 times table is super simple and fun. Let's start our journey!

1. What is the 10 Times Table?

The 10 times table is a special way to multiply numbers by 10. Every time we multiply by 10, the number increases by one place! Here's how it works:

10 x 1 = 10
10 x 2 = 20
10 x 3 = 30
10 x 4 = 40
10 x 5 = 50
10 x 6 = 60
10 x 7 = 70
10 x 8 = 80
10 x 9 = 90
10 x 10 = 100

Easy, right? Now let's see how we can learn it in a fun way!

2. How to Learn the 10 Times Table in a Fun Way

Learning the 10 times table can be an adventure full of games and creativity. Here are some super fun ideas!

2.1 The 10 Times Table Song

Singing is a fantastic way to remember! Try singing this little song to the tune of a song you like, such as a popular tune or a lullaby. Here's an example:

? Ten times one is ten, you already know!
Ten times two is twenty, let's all shout, go!
Ten times three is thirty, what a thrill,
Ten times four is forty, join the fun, we will! ?

Keep singing all the way to 10 x 10 = 100. Every time you sing, you'll remember the numbers more easily!

2.2 The 10 Jumps Game

Ready to jump? This game will have you moving like a kangaroo! Every time you say a multiplication result, take a jump!

Jump for 10 x 1 = 10.
Take another jump for 10 x 2 = 20.
Keep jumping all the way to 10 x 10 = 100!

You can do this outside or even indoors, as long as you have enough space to jump. Every jump will help you remember the results!

2.3 The 10 Candles Game

Imagine you are a baker preparing a big cake for a party! Every time you multiply by 10, light a "candle" (you can draw candles on a piece of paper).

For 10 x 1 = 10, draw 1 candle.
For 10 x 2 = 20, draw 2 candles.
And so on until 10 x 10 = 100.

Every time you light a candle, say the result out loud. By the end, you'll have a big party with lots of candles!

2.4 The 10 Times Table Puzzle Game

Draw a big puzzle on a piece of paper. Each puzzle piece represents a multiplication from the 10 times table. For example:

Draw one piece with 10 x 1 = 10.
Another with 10 x 2 = 20.
And so on up to 10 x 10 = 100.

Cut out the pieces and mix them up. Now, your goal is to put the puzzle back together by matching the results with the correct pieces! This will help you remember the numbers while having fun.

2.5 The Magical 10 Ladder

Take a sheet of paper and draw a ladder with 10 steps. On each step, write one of the results from the 10 times table:

On the first step, write 10.
On the second step, write 20.
On the third step, write 30.

Every time you climb a step, say the result out loud. When you reach the top, you'll have reviewed the entire 10 times table!

2.6 The Magic Dice Game

For this game, take two dice. Roll the dice and multiply the number you get by 10. For example:

If you roll a 5, you must say 10 x 5 = 50.
If you roll a 3, say 10 x 3 = 30.

You can challenge a friend to see who can answer the fastest!

2.7 The Table Decoration of 10

Imagine you are decorating a big table for a party. Take a sheet of paper and draw a table with 10 plates. On each plate, write a result from the 10 times table:

First plate: 10
Second plate: 20
And so on until 100.

Color each plate with different colors. Every time you look at your colorful table, you'll easily remember the 10 times table!

3. Review Every Day with Joy

Now that you know so many games and tricks, it's important to review the 10 times table every day. Here are some ideas:

Jump and sing the 10 times table song while doing your homework.
Play The 10 Candles Game with a friend.
Do the 10 Puzzle and challenge yourself to complete it faster each time.

Reviewing the 10 times table a little each day will help you remember it effortlessly. You'll become a real champion!

Conclusion

Learning the 10 times table is like a journey full of adventures and fun! With songs, games, jumps, and colors, this times table will become one of your favorites. Have fun, little number explorer, and remember: math is a great adventure!

Index

1. Introduction to the World of Multiplication Tables pg.5

Chapter 1: Learning the 1 Times Table – A Super Easy and Fun Game! pg.7

Chapter 2
Learning the 2 Times Table: A Fun Adventure! pg.11

Chapter 3:
Learning the 3 Times Table: A Game of Magic and Numbers! pg.15

Chapter 4: Learning the 4 Times Table: A Team Game! pg.19

Chapter 5: Learning the 5 Times Table: A Fun Journey! pg.24

Chapter 6: Learning the 6 Times Table: A Fun Number Adventure! pg.29

Chapter 7: Learning the 7 Times Table: A Super Fun Adventure! pg.34

Chapter 8: Learning the 8 Times Table: A Magical Journey Through Numbers! pg.39

Chapter 9: Learning the 9 Times Table: A Fantastic Adventure Among Numbers! pg.45

Chapter 10: Learning the 10 Times Table Easily! pg.50

www.ingramcontent.com/pod-product-compliance
Lightning Source LLC
Chambersburg PA
CBHW070418230526
45471CB00006B/2861